The Naked Darwinist

The Naked Darwinist

Elaine Morgan

During the last few years, when I have found myself in the company of distinguished biologists, evolutionary theorists, paleoanthropologists and other experts, I have often asked them just to tell me, please, exactly why Elaine Morgan must be wrong about the aquatic theory. I haven't yet had a reply worth mentioning, aside from those who admit, with a twinkle in their eyes, that they have also wondered the same thing.

— Daniel Dennett

Eildon Press

© Elaine Morgan 2008

Eildon Press

Printed and bound by Antony Rowe Ltd, Eastbourne

ISBN 09525620 30

Related books by Elaine Morgan

The Descent of Woman, 1972, Souvenir Press

The Aquatic Ape, 1982, Souvenir Press

The Scars of Evolution, 1990, Souvenir Press

The Descent of the Child, 1994, Souvenir Press

The Aquatic Ape Hypothesis, 1997, Souvenir Press

Contents

Part One: The story so far

Don't ask	3
Raymond Dart	8
Alister Hardy	14
The "feminist tirade"	18
Just ignore it	23
The brighter side	27
Contact	32
Other sources	37
What the fossil record proved	42
The aftermath	48

Part Two: Objections and replies

Objections and replies	55

Part Three: What makes human bodies special

We walk upright	63
We have a displaced larynx	71
We can talk	80
We put on weight	83
We are furless	87

Part Four: Present and Future

A fairy tale	101
Where we stand today	102
The road ahead	107

Part One
The Story so Far

Don't ask

IMAGINE you are a student revising for an exam on human evolution and wondering what questions you might be asked. It suddenly occurs to you that you cannot remember why human beings lost their body hair. It sounds just the kind of topic on which they might give you a quote and then say: "Discuss" - and the answer has gone right out of your head. You cannot even remember the point being raised, and the exam is tomorrow. What can you do?

You can relax. That question will not appear on the exam paper. It never does. It is as if there was an unspoken agreement between the people teaching this subject and their students, to the effect that if you don't ask them that question, they will never ask it of you either.

If you think I am exaggerating, take a look at the reference books on human evolution which have to be compiled every decade or so to provide students with the most up-to-date information. In 1992 *The Cambridge Encyclopedia of Human Evolution* was published. It was over five hundred pages long, and gave comprehensive coverage of all other aspects of the subject, but it nowhere referred to the fact that humans have lost their body hair. An accidental oversight, perhaps? In 2004 a completely new reference book appeared entitled *Principles of Human Evolution,* compiled from scratch by different editors. It too was 500 pages long. It made the same omission.

Elaine Morgan

If the entire contents of these books could be beamed up to one of those mythical life-forms on some distant planet, the aliens would be left with the mistaken idea that we are just as furry as our nearest relatives.

On the face of it, it seems a bit odd. If you ask anyone with an unsophisticated mind - say a nine-year old child - to name differences between a man and a chimpanzee, the list will certainly contain, somewhere in the top five, the facts that a man walks upright and a chimp goes on all fours, that a man can talk and a chimp can not, and that a chimp is hairy but a man is not. You cannot fail to notice that last one. It hits you in the eye. Yet somewhere on the long road between ignorance and specialisation, its perceived importance in the evolutionary story dwindles away until finally it sinks without trace.

You may think this is a fuss about nothing. Assuming that all the other features distinguishing man from the other apes have been explained, how much does it matter if we cannot quite account for this one little aspect of it? But that is not quite how things stand. There is no agreed explanation for *any* of the anatomical features distinguishing humans from their nearest relatives. Not one. Take bipedalism for example. The text books will give you a list of possible reasons that have been suggested as to why *Homo sapiens* is the only mammal that walks upright. None of these is convincing enough to be acclaimed as the right answer, but they are all worth bearing in mind. The only difference between the attitude to this and the attitude to hairlessness is that in the latter case they don't offer you the list.

There is one other question that you are not encouraged to ask. In its simplest terms, it is "Why only us?"

We are told that all these changes happened to our species because the ancestors of chimps and gorillas stayed in the trees and our ancestors came down to the ground. But they were not the only primates to leave the trees. Think of all the baboons and the geladas and the vervets and the patas monkeys and the Barbary apes and the Hanuman langurs. If we rose up on two legs because it was faster or more efficient or enabled us to see further, why didn't any of the others do the same? Not even one of them? If we became naked because of over-heating in the chase, why didn't all the animals we were running after and all the ones we were running away from become naked as well? They were moving just as fast under the same hot sun. If we became the sweatiest animal in the world in order to cool down, why did the camel protect itself against over-heating by hanging on to its woolly coat and reducing its sweating to the barest minimum? Why did our ancestors so often respond to exactly the same problems that confronted other animals by adopting diametrically different strategies? There may be good answers but I don't hear them, and it is not because I have not been listening.

Elaine Morgan

There is a mystery here. The people who write the texts have devoted their lives to the pursuit of truth and to throwing light on how and why the human race emerged. They are professional in the highest degree, and have earned the respect of their peers. The books are intensively researched, lucidly presented, graphically illustrated, meticulously documented and referenced, and bang up to date. In most instances where questions remain unsolved, that fact is duly noted, as the ethics of scientific enquiry dictate.

So what is happening? I do not for a second suspect that they sat down and conferred about the nakedness of the naked ape, and consciously decided that it would be more prudent to shut their eyes to it. It is more mysterious than that. I think that throughout all the months of hard labour they put in, the issue never once crossed their minds. It had somehow been collectively blotted out. In a different context it might be said that they were in denial about it. What then goes on in their minds? How do we account for the strange behaviour of the subspecies *Homo anthropologicus*?

A recent weblog on pseudoscience posed the pertinent question: "If the Aquatic Ape Theory explains so much, why do the majority of anthropologists not subscribe to it?" What follows is an attempt to answer that question. But I have to remind myself of one thing before I begin: if the answer is to be of any value, I must be able to feel at all times that in their position I might be behaving in exactly the same way. Otherwise it would not be an explanation. It would be a piece of pure paranoia.

Elaine Morgan

Raymond Dart

IF YOU BELIEVE, as all Darwinists do, that our species once shared an ancestor with African apes, there are two ways of seeking evidence for that belief. One is by means of physical anthropology - comparing human anatomy with that of the apes. The other is through the study of palaeontology - examining the fossilised remains of creatures that might have been our ancestors. Darwin used the first method because he had no choice, since hardly any fossils had been discovered in his time. Nobody even knew in which continent to start looking for them, though Darwin thought Africa would be the likeliest place.

Much later, in 1924, a young Australian called Raymond Dart, who had studied in England before moving to South Africa, announced the discovery of the fossilised skull of a young creature with, he claimed, some human-like characteristics. The top specialists in the field, Wilfred le Gros Clark and Sir Arthur Keith, looked at photographs of the skull. It was slightly more rounded than that of the average chimp, but seemed much too small to be pre-human. They remembered teaching Dart when he was still wet behind the ears, and he hadn't struck them as the kind of student who was going to rock the scientific world to its foundations. Quite understandably they said it was a chimpanzee's skull, and they went on saying it for 23 years. Then Le Gros Clark proved his open-mindedness by going out to Africa and holding the skull in his hands. He changed his mind and said Dart had been right. Later Sir Arthur Keith did the same thing and reached the same conclusion: that this little skull of the "Taung baby", was indeed a long sought "missing link." And Africa was the place to look for more of them.

Dart did not wait for this validation before publishing his discovery and what he thought it meant. Some earlier believers in evolution had had a vague vision of a naked ape with an unusually large head swinging through the treetops. We don't often remember that, but Rudyard Kipling's *Jungle Book* and the Tarzan stories of his cousin Edgar Rice Burroughs were written at that period and they reflect that belief. Dart is usually given the credit for replacing this picture with the image of a savannah ape. As recently pointed out by Renato Bender and Phillip Tobias, he was not the first to visualise our first ancestor moving from the trees to the plains. One of the earliest to think in those terms was Lamarck. But Dart played a major part in fixing that vision in the public imagination. He described how those early ancestors in their new habitat found they could no longer live on the fruits of the forest and were forced to become hunters. He drew bloodthirsty word-pictures of how they must have used animal bones to fashion weapons for killing prey and ripping up the carcasses, and possibly for killing each other too. It was an exciting and popular idea, even for people who had not previously shown much interest in anthropology.

Once the Taung skull received the official seal of approval, the hunt for fossils was stepped up. More specimens came to light, and fossil hunting became the new exciting field, the growth area of evolutionary research. It involved travel and adventure, and if you actually found a proto-human fossil it brought fame and glamour. You would be photographed standing on a slope of barren ground under a sweltering sun, and there would be newspaper articles with the kind of headlines that still occur quite frequently: "New fossil find forces re-think on human origins." You could not get publicity like that by studying human anatomy, because all the big discoveries, like the circulation of the blood, had been made long ago. Which career would you have chosen? You would probably have opted to become a PA - a palaeoanthropologist.

So the research into our evolutionary past was split into different specialities which did not communicate with one another much, if at all. For most people the science of human evolution came to mean discovering and examining fossils. You can see how that might influence thinking on the subject. If you are repeatedly confronted with carcases of apes and humans to dissect, you cannot help being reminded that one species is naked and the other is hairy, even before making the first incision. But if you are repeatedly asked to examine and describe bones and teeth, you tend to think that understanding bones and teeth is the only highroad to the truth about the human race.

Dart's vision was gradually promoted from being a young man's fanciful idea into a scientific orthodoxy. We had long known that at one period human ancestors had roamed the African savannah and hunted game - their weapons and cave paintings testified to that. We were now informed that this environment was the factor that had split us from the apes in the first place, and started us on the path that led to what we are today. The scientists with the clearest understanding of how natural selection works were the most enthusiastic. They felt the savannah hypothesis represented a big step forward in evolutionary thinking. And they were right to feel that.

In the early days of Darwinism, most of the debate had been over whether or not we descended from an ape-like ancestor. Until that basic battle had been won, nobody concentrated much on why these changes took place. There was a wide-spread belief that humans are at the top of an evolutionary tree, as if some life form or other was predestined to arrive at this pinnacle of intelligence and efficiency, and it so happened that on planet Earth our species had won that race. T. H. Huxley could pack the Albert Hall with audiences listening with rapture to how ancestral humans became top species by aspiring to this eminence and overcoming all obstacles in their path. Even today in science fiction, most alien life-forms from distant planets infallibly end up intelligent, verbalising, and with an erect form of locomotion.

But that is not really what Darwinism implies. Natural selection operates by means of the selective death of the relatively unfit - and that means unfit for the conditions in which they are currently living. They do not aspire to evolve to suit tomorrow's conditions. There is no road-map and no destination. For the scientists with the clearest understanding of that, Dart's paradigm felt like an advance to a more realistic way of approaching the problem. If our ancestors found themselves in a new and quite different environment - like the savannah - then of course they would begin to change. It all made a lot more sense than Tarzan in the trees. They were now in a position to ask new questions, and think more constructively about what the new answers might be.

It is easy to see how intellectually satisfying it was to contemplate the ape that moved out onto the savannah, and how it inspired people to look for new proofs that it had happened. The fact that Dart himself had initially had to fight against doubters and deriders only deepened the commitment to his concept among young scientists eager to make their mark.

One great thing about the new scenario was that it was so easy to communicate the essence of it even to laymen. Books were published outlining it for the benefit of the general reader, and proved immensely popular. They explained how the species that remained in the shrinking forest dwindled in numbers - and are still dwindling. But our own ancestors struck out, left the trees, and with strong pressure on them to increase their prey-killing prowess, vital changes began to take place. They became more upright - fast, better runners. Their hands became freed from locomotion duties - strong, efficient weapon-holders. Their brains became more complex - brighter, quicker decision-makers. And at the height of the chase they became overheated and sweated profusely; and that caused them to become naked, to allow the heat to evaporate more readily and cool them down.

All the pieces were falling into place. You can imagine how intensely irritating it would have been at that point if any scientist had broken ranks and spoken up to suggest that maybe they had got it all wrong. As it happened, throughout the thirties and the forties and the fifties, no-one showed any signs of doing that. There was only one man in England who was privately convinced they had got it all wrong, and he was too prudent to voice his objections out loud.

Elaine Morgan

Alister Hardy

IN 1930 ALISTER HARDY, a young Oxford marine biologist, read Prof. F. Wood Jones's comment that one striking difference between apes and ourselves is the layer of subcutaneous fat adhering to the human skin. Because his speciality was the ocean, Hardy was instantly reminded of the similar fat layer found in aquatic mammals like dolphins, seals, and manatees, often in conjunction with a naked skin - another anomalous human feature. Could it be that our ancestors had been more aquatic at some time in the past? The same wild idea had previously occurred to one or two earlier academics - notably Max von Westenhofer in Germany. His published observations on the subject were more extensive than Hardy's ever were, and his prior claim is perhaps unfairly neglected. But he found no support for the idea. In later editions of his book he omitted to refer to it and it was quickly forgotten. Hardy conceived the idea independently, the way Alfred Russell Wallace conceived of natural selection, and he never did discover that he had been forestalled.

For thirty years he kept quiet about it. He was at home in the academic community, he knew the ropes, and he was ambitious. In his old age in a filmed interview he said frankly: "I wanted to be a professor. I wanted to be a Fellow of the Royal Society." There were things you did not do if you hoped for those rewards.

In 1960, he had achieved both of those objectives and a knighthood into the bargain, so he slightly relaxed his vigilance and confided his thoughts, just for fun, to a small lay audience - a sub aqua club in Brighton. An enterprising local reporter was present and the national Sunday newspapers carried banner headlines: "Oxford professor says man a sea ape."

Le Gros Clark - he was still around - was understandably livid. He rang up and issued a stern edict: "Alister? Never do that again!" Anyone in his place would have been liable to feel the same shock and dismay. Alister, not foreseeing the publicity, had not consulted his colleagues, had not submitted his ideas to peer review, was trespassing on other people's preserves because he was not and never had been an anthropologist, and he had exposed Oxford science to public ridicule by airing such a bizarre and childish notion. There is no evidence that anyone tried to discuss his idea with him or present him with arguments about why he must be wrong. They simply told him to shut up.

The whole affair was quickly smothered. Alister reserved the right to contribute one article to *The New Scientist* and give one talk on the BBC's Third Programme, to correct some garbled versions of what he had actually said. His colleagues then forgave him, comparatively few people ever heard of his gaffe, and for the time being the matter was closed. After his death, when a memorial service was held to honour his memory, no mention was made of this incident in his life, as if it had been a discreditable aberration on his part which it was kinder to forget.

As with an outbreak of foot-and-mouth disease, there was a sigh of relief that this particular heresy seemed to have been contained. Hopes were high that it had been stamped out in time to prevent anybody else from being infected by it. I would certainly never have heard of it but for a reference to it in Desmond Morris's *The Naked Ape*. Desmond, who knew and admired Alister, gave a brief account of the theory, but concluded that if there had been an aquatic interlude, its effects were probably minimal. The big story, the true story, was still the exodus onto the plains.

But when I read that passage, it came as a revelation. To me it made more sense than anything else in the book. This was at the beginning of the seventies, and I had started writing a book of my own, with a feminist message. Its theme was that all the current books on human evolution were written from a male standpoint. Some of the changes allegedly designed to make the male a better hunter would have had unfortunate consequences for his mate and her offspring. A female primate in the middle of an African plain would not be better equipped to find food, or to escape from predators, by becoming naked and bipedal. That would involve having to carry everywhere in her arms, while foraging or running away, a large and particularly helpless slow-growing infant that could no longer travel astride her back, nor cling to her fur so as to leave her own hands free. I preferred the aquatic hypothesis. I wrote to Sir Alister, asking for and receiving permission to follow up the suggestion he had made.

Elaine Morgan

The "feminist tirade"

THAT BOOK, *The Descent of Woman*, became a best-seller, and in the United States it was a Book of the Month selection. I became a minor celebrity, ferried around America coast-to-coast for a couple of weeks, appearing on the top chat shows. The book was a hit with the public - or with at least half of it. It was understandably disliked by some of the good ol' boys who couldn't stand sassy women. But what about the scientists? I had never expected it to be acclaimed by them. I remembered how they reacted to Alister, and my case was far worse. I had no qualifications for presuming to tackle that subject, the tone of the book was combative and flippant, the "research" sketchy and superficial, and the theme blatantly politically motivated. Their reaction was to ignore it. Insofar as it affected them at all, it was only in two respects. One was about pronouns and the other was about the penis.

I had stressed the difficulties caused by the fact that in English "man" has two meanings. Sometimes it means the human race and sometimes only part of the human race. Referring to our first ancestor as "he" could lead to the subconscious assumption that evolutionary change in the species occurs to improve the survival chances of the male hunter, regardless of any knock-on effect it might have on his mate and offspring and their chances of survival. That would be an unsafe assumption.

I was of course not the first or only one to say that, but the book was making a noise, and it appeared at a time when Women's Liberation was just beginning to get off the ground. Some of them were marching around the campuses with banners, and there was a period when some academics became quite nervous about their pronouns. How could they be expected to write: "When our first ancestor left the forest, he/she found it was necessary for him/her to change his/her mode of subsistence"? You can see how exasperating it was to have to seek ways of avoiding that kind of sentence, and to have to read through every paper and weed out sexist pronouns before submitting it. The final effect however was benign. The simplest solution was to make greater use of "they" and "their" and that created a much more realistic picture of events. No man is an island, and no woman either.

The penis thing was relatively very minor. There had been a dispute over why the penis in humans is so much larger than in other apes. Some claimed it was in order to allure females, others that it was to impress and intimidate other males. I made the purely utilitarian observation that bipedalism had made the vagina relatively inaccessible, to the point where a gorilla-sized penis would be totally useless. This was accepted, with the proviso that it would have happened for *whatever* reason the species became bipedal, so it could not - repeat *not* - be advanced as an argument for the Aquatic Theory. Fair enough.

In all other respects, I braced myself for the possibility that the aquatic sections would be attacked and perhaps demolished. I expected some scholarly figure to be invited into a studio where he would say "The points the author fails to take into consideration are (a) and (b) and (c). These facts alone render her idea unacceptable." That would have left me wiser and not appreciably sadder. I had not invested much intellectual capital in AAT at that time. At least I would have learned something.

The response I had not foreseen was total silence. But in respect of the aquatic theme that was what I got from them - and with few exceptions still get. That kind of silence is a virtually unbeatable strategy. My first assumption was that in their view there are people who qualify for being replied to, and people who do not. In the old days of jousting and duelling, the rules were well understood. It was a point of honour for a gentleman never to refuse to defend his own good name or that of his beloved in single combat, against any other gentleman who threw down a gauntlet. But it was equally a point of honour to decline to tangle with anyone who was not a gentleman. Some remnant of that attitude seemed to be still in operation.

That was a silly and ignorant interpretation of the silence. A moment's thought should have told me that the feudal comparison is quite untenable. Every profession has a duty to establish its own standards and require its members to abide by them and show some solidarity. That is even truer of this subject than of "harder" sciences, because you are talking about the human race, so people take it personally and think they have a right to butt in. And then there were the creationists, also on the march. Arguing with people who start from such totally different premises as the creationists is a waste of time. It gets you nowhere. A line has to be drawn. The same, you may feel, goes for pseudo-scientific nutters like believers in flying saucers, and aquatic apes, and stuff like that. All the ideas being handed down to the new generation have been exhaustively probed and pondered over for years, before being admitted to the canon. If they had to recap all the evidence whenever anybody piped up "I don't get it", they would never get any work done.

If at Oxford I had read Anthropology instead of English Literature, would I have shared that attitude? More likely than not, yes. Many of the leaders in the field were and are undeniably brilliant. Judging by the ones I have been privileged to meet, the profession attracts more than its share of charismatic figures. I would have been dazzled by them. Besides, loyalty is a quality held in high esteem in all areas of society. If you belong to a category of people and hear that category being criticised - whether it is differentiated by race or sex or class or occupation or any other criterion - there is a basic human tendency to close ranks and defend it against outsiders.

Just ignore it

AS TIME WENT BY it became clear that nobody was going to attack AAT: Aquatic Ape Theory. It was more effective, as well as easier, to say nothing. Nobody got the impression that the case for an alternative to Savannah Theory had gone unanswered. Everyone assumed it had been examined and disproved so promptly and decisively that the actual text of the refutation had escaped their notice. In 1997 a columnist in *New Scientist* happened to pick up a copy of *The Descent of Woman* and read it for the first time. He commented: "One hears that Ms Morgan got a fearful towsing from other ethologists ... and is now relegated to the bleachers to ponder the error of her ways." Yes, one did hear that and one still hears it. But the towsing must have been administered in my absence (to an effigy, perhaps?) and behind closed doors. We are still repeatedly assured that the idea has been officially examined and found wanting, but I have never been able to find out who conducted this enquiry, or where, or when.

For more than twenty years any attempt to question the consensus was blocked by two objections so obvious and so massive that they did not even need to be voiced. For one thing, there was no market for new theories because virtually everybody was perfectly happy with the one they had. The move to the savannah had explained much, and would, it was assumed, ultimately explain everything. Secondly, there was the intellectual high ground occupied by the PAs - the palaeoanthropologists. They alone knew, as the saying goes, where the bodies were buried. They knew that if you wanted to find fossils of pre-human creatures you had to go out onto the savannah to look for them. And lo! there they were. It was a fact, and you cannot argue with facts.

It must have been around then that the question "Why naked?" began to be dropped from the agenda by common consent. There was no conscious decision and no concealed motive. Any scientist worth his salt concentrates on questions he has some hope of answering, and particularly on those where new and relevant evidence has come to light. Newly discovered fossils of skulls and teeth and femurs and foot bones could be examined, measured, and written up. Compared to that it would be a frivolous waste of time to dream up scenarios about loss of body hair when hair does not fossilise.

After ten years of the silence, I wrote a book entitled *The Aquatic Ape*. No jokes this time, no sex, no feminism. Just sober prose - and it did have one good effect. There had been a persistent rumour that Hardy had only written his two articles as a mischievous kind of academic joke, which I alone had been humourless enough to take seriously. He wrote a preface to *The Aquatic Ape* which at least knocked that idea on the head, and that was fine. So what was the scientists' response to it? None. Why should I have expected anything else? If I had been in their place would I have behaved any differently? Of course not. We all behave like that. I have never bought a book about spiritualism or UFOs in order to weigh up carefully the arguments for and against. I feel convinced that they must be tosh, just as others were convinced that mine must be tosh.

In 1990 I wrote *The Scars of Evolution*. It was much better than its predecessors. A full-page review in the British Medical Journal acclaimed it: "Elaine Morgan seems to have succeeded where the professionals have failed." What did the professionals say? The *New Scientist* review struck a familiar note: "There is hardly a mention of the large body of evidence that refutes it, or of the many reputable scientists who remain unconvinced by it."

Douglas Adams advised me to try the Internet and I took his advice. When I joined a group discussing evolution, I was met with a torrent of scorn and hostility and urged to get out of their air-space and go back where I came from.

Since I seem to be describing a process of banging my head against a brick wall, you may reasonably ask why I kept on doing it. Why not chuck it in and return to writing for television where I had been doing very well, winning Baftas and other awards? The nearest I can get to explaining my persistence is to borrow an

Elaine Morgan

odd phrase Darwin once used to convey why he was so anxious about how *The Origin of Species* would be received - not for himself, he wrote, but "for the subject-sake."

The brighter side

However there is a world outside academia, and inside academia there are faculties other than palaeoanthropology. In principle there is no reason why PAs and anatomists should not work co-operatively together, each contributing their own kind of data to the final picture that would emerge. But one or two of the people who study soft tissues rather than bones were inclined to register doubts about the savannah scenario which had proved so inspiring and fruitful. They pointed out that nakedness under a hot sun does not cool an animal. If you shave the hair off its back, its core temperature goes up, not down. They noted that humans sweat more profusely than any other animal on the planet - the very opposite of what would be adaptive in a habitat where water and salt were scarce resources. They said fast-running predators and prey do not normally accumulate deposits of fat, which would only slow them down. In other words, there were still a few question marks.

Then there were the swimming babies. One of the most devastating arguments against an aquatic interlude was that all the babies would have promptly drowned. But suddenly there were headlines reporting that if they are introduced to water early enough, they thoroughly enjoy it. And all that fat that makes them look so different from other primate infants helps to keep them afloat.

In 1981 an American geologist Leon P. La Lumiere drew attention to an area near the Red Sea called Afar which millions of years ago had been subject to massive sea-flooding and turned into the Sea of Afar. He suggested that some apes in that previously forested region could have been trapped on shrinking islands of high ground and been forced to turn to marine food sources. The idea was disregarded, because all the oldest fossils had been found in South Africa. But then Lucy was discovered - at that time the oldest of fossils as well as the most famous. Where? In Afar. In hindsight that does not prove anything at all - fossils are now turning up all over the place, and I am no longer wedded to the idea that the immersion was in salt water rather than fresh. But at that stage of the game it felt encouraging. It showed we all still had a lot to learn.

After a time other people began to voice support. First Carl Sauer published a paper. in 1983. In 1985 Marc Verhaegen entered the arena with a paper in *Medical Hypotheses* entitled *"The Aquatic Ape Theory - Evidence and a Possible Scenario."* Marc soon proved to be quite as strongly motivated as I was to overcome the blanket resistance to the idea, and, as a doctor, he was able to supply more anatomical data than I had managed to collect at that time. Since then he has been one of the most tireless advocates of an aquatic influence on human evolution, and has constructed scenarios of his own concerning timing, type of habitat, and the shape of the family tree.

For a period of five years, New Scientist accepted articles from me about different aspects of the theory, at the rate of about one a year, until they received complaints that this was an unacceptably one-sided policy since they never printed articles critical of it. They said that was a fair criticism and returned my latest piece saying they could not accept any more unless and until some opponents of the theory chose to voice their views. I knew that would never happen - they had learned by experience that silence is golden. I rang up and offered to write some stringent attacks on AAT myself, under a pseudonym, rather than let the whole debate be effectively gagged. They thought not. I cannot blame them for not collaborating in such a subterfuge. It was many years ago and the ban has long since been lifted or forgotten.

Michael Crawford, head of the Institute of Brain Chemistry and Human Nutrition in London, was researching the nutritional needs of brain tissue and noted that it depends for healthy growth on a particular balance between Omega 3 and Omega 6 fatty acids, precisely the balance that is found in the sea food chain. He speculated that a switch to marine food resources might have made possible the rapid increase in brain size in *Homo,* and his work created great interest among evolutionists as well as nutritionists. His influential book *The Driving Force* induced many people to look again at Hardy's ideas.

There was an upside to the Internet, too. Outside of cyberspace it has always been hard to find what the opposition is thinking, other than by chance. For example, one London journalist wrote to Stephen Jay Gould asking for his verdict and sent me his reply to get my response, despite the fact that it was labelled "off the record." It said: "I do regard the theory as so entirely disproven that its maintenance can only be labelled as something close to crackpot. Every vertebrate lineage - really every single one - that has ever returned to even partial life in water (as in otters) lose strength for walking in their legs." Now that was a solid argument. I did not find it conclusive (a flamingo is part of a vertebrate lineage and its legs grew longer) but it was certainly worth thinking about. If there was more where that came from, why was it meant to be off the record? What was secret about it? Why did nobody assemble it and publish it? It would not have taken long to write and there would have been a healthy market for it, at a time when professors were muttering that this issue was becoming a nuisance to them.

One of the joys of the Internet is that whatever its faults, they do not include a determination to remain shtum. All the cards were on the table. And what did they amount to? It became perfectly clear that the reason for not believing it is that people do not believe it. I was told over and over again that nobody - but nobody - believes it. Why could I not accept that? Why could we not settle this thing once and for all by a show of hands? I grew happier. I said to myself "They haven't got a case." There may well be a case. It may be a strong case. But no one seems able to convey it to the rank and file in such a way that they can assimilate it and relay it to other people.

Elaine Morgan

Contact

IT WOULD be wrong to give the impression that I never came into contact with any scientists. I was giving myself a crash course in evolutionary studies. Although I lived a long way from London I found I could avail myself of most of the facilities of the British Library by post, without actually going up there and taking a seat in the famous Reading Room that it then still occupied, even if it took a little longer that way. Sometimes too I would go up and attend a public lecture if it sounded particularly inviting. But also over the years I have had official invitations to attend and address seminars on various campuses - a score or more altogether, from half a dozen different countries and on both sides of the Atlantic. I was invariably treated with the utmost courtesy, and allowed to say whatever I liked.

These occasions were always stimulating and enjoyable, and there was only one drawback. From my angle they were not informative. Following the presentation there would be a session of questions from the students, which I welcomed. It helped me to understand what topics might be pre-occupying them. But there was no way of knowing what answers might be given to those questions or any others by their mentors after I had gone. They were not the ones in the dock. Nobody offered to *debate* the issue. You may ask, if I had been in their position, would I have organised matters in that way? You bet your sweet life I would! I have done a bit of teaching. Not much, but quite enough to know that it is a tough job on any level. The last thing you need is some outsider coming in and questioning you in front of your own class.

There had been one egregious exception to this arrangement. By 1987 there were enough people who were sympathetic to the aquatic idea to make it possible to contemplate a meeting of minds. A conference was jointly organised by the European Sociobiological Society and the Dutch Association of Physical Anthropology. It took place in Valkenburg, and its raison d'etre was explained as follows: "Humans are in various aspects so different from other animals - including other primates - that, according to Darwinian theory, we could only have evolved when our earliest hominid ancestors occupied a specific niche, quite different from that of the other contemporary primates." The question to be addressed was: What niche?

It was just what I had always dreamed of. There was very little agreement but there was an open-minded and entirely good-tempered exchange of views. In many international conferences there are some sparsely-attended sessions because delegates have taken the opportunity to do some shopping or wander off to the see the sights, or else they nod off in the back row. In Valkenburg there were queues outside the lecture room each morning waiting for the doors to be opened, and nobody fell asleep. It was a well-balanced programme, and the discussions ranged over a wide range of topics, including primate behaviour, marine ecosystems, geophysical events, comparisons between fossil hominids of different dates and living species, the cultural anthropology of present-day tribes of hunters and of fishers, and physiological differences between apes and humans. Marc Verhaegen and Leon la Lumiere were present, together with newer supporters including Derek Ellis, Erika Schagatay, and Karl-Erich Fichtelius, who made then and continued to make valuable contributions to the debate.

Rarely was there any disagreement about the facts, since both sides were relying on well-documented academic sources. If anyone did happen to be under a misconception, they were put right and ended up better informed. That worked in both directions. For example, I had thought the presence of a hymen in human females might be significant, since apes do not have it and many aquatic mammals do. When I discovered it is also present in horses, I dropped that argument. On the other side, one contributor argued that we are uniquely ill-adapted for life in water because we are only protected from the air entering our air passages by the pressure of air in our lungs. In fact we have evolved a moveable velum which serves that purpose.

I got the impression that everybody left Valkenburg comfortably convinced that they had put up a good show. That should surprise nobody. Even after the famous Darwin debate in Oxford in 1860, which T.H. Huxley so often recalled as a resounding defeat for Bishop Wilberforce, there is documentary evidence that the bishop felt equally triumphant, and was congratulated by his friends for having wiped the floor with his challenger. Unless I was deluding myself - which is always on the cards - what happened in Valkenburg and did *not* happen in 1860 was a slight, a very slight, narrowing of the gap, a decrease in mutual suspicion, a little more tolerance all round. There seemed even to be a chance that these two lines of thinking are not irrevocably parallel, but might succeed in converging, at some point on this side of infinity. One statement made was that the niche occupied by a pre-human ape could only have been at most "semi-aquatic". That is tenable. We could build bridges with that.

One point I think is beyond question. If the topic of debate had been visitors from outer space, such a conference could not have been sustained over three days and on that level of rationality. Later an account of the proceedings was published under the title of *The Aquatic Ape: Fact or Fiction?* It is by now mainly of historical interest because in this field new facts accumulate rapidly and perceptions keep changing. But the summing up by the editors faithfully reflected the overall majority opinion that the aquatic case was very interesting - but unproven. After all, there were the grasslands. There was the fossil evidence. There was the mountain of theses and speculations that had been erected on the basis of the Savannah Theory (and in those days nobody had the slightest qualms about calling it the Savannah Theory.) It had served them and their predecessors supremely well for over half a century. It was not going to be overturned by its failure to resolve a few little perceived paradoxes. But it also reflected one other conclusion. One speaker commented that in the minds of many scientists AAT had apparently been "miscategorised" That seems the right word for it.

Other Sources

THAT EXPERIMENT was never repeated. I had to look for other ways of keeping up with what the official stance might be. One other possible source should have been the major professional journals. Unfortunately they continued to behave as if they were oblivious of the existence of any alternative to the savannah paradigm. They accepted no paper outlining the case for the Aquatic Theory. They also accepted no paper denouncing it - perhaps because that might have seemed to sanction the right of reply. But here again there was one exception to the rule. John Langdon figured that he might be allowed to take a swipe at it if he treated it as just one random example that he happened to light on, to illustrate a much larger and more philosophical theme. His paper appeared in the *Journal of Human Evolution* in 1997.

He was unfortunate in one respect. My last book *The Aquatic Ape Hypothesis* was published while his paper was still in press, and that meant that some of his criticisms were already out of date. Not that dates worried him: some of the things he denounced were things I wrote in that first fine careless rapture of 1972. Few scientists in this field would expect to be attacked for statements they made that long ago, even if they had been fully qualified when they wrote them.

But he wished primarily to deal with categories. He was anxious to relegate AAT to a specific category of ideas which scientists are justified in rejecting without examining them or replying to them. Others have tried to do the same thing, so far without success. Langdon offered the term "Umbrella Hypothesis" to describe "an idea that overspreads and appears to resolve many scientific questions" - especially if it is easily communicated and understood by the general public.

If all such ideas are to be rejected out of hand, then bang goes Natural Selection, the most famous of all umbrella theories. He needed a category that would exempt Darwin but stop Hardy in his tracks, in company with creationists, homeopaths, and extra-terrestrials, so he resorted to the word heterodox. Heterodox ideas, he told us, feed on "a suspicion of and rebellion against established science and other authority", and have a special appeal for "peripheral segments" of society. He cites African Americans, and AIDS victims (code-word at that time for homosexuals), and he detected in my first book "the passion of embittered and victimised feminism." (Embittered? I wouldn't have said that. Quite cheerful, really.)

The message is clear: he is saying that those who challenge the conventional scenario are not the kind of people you would want your sister to marry, and that their attitude to scientific questions is contaminated by their political leanings. It is perfectly true that people's scientific ideas are sometimes influenced by their political views and I would never claim to be immune from that tendency. If we were discussing for example Nature v. Nurture, I am sure my attitude would be no less (and no more) contaminated by social background and life experiences than Professor Langdon's.

But the aquatic theme is uncontaminated. It is innocent of any political implications. It was pure chance that I was writing a feminist book when I first heard about it. There have been people among its supporters - including that dear and gentle man Sir Alister himself - whom nobody in his senses would include in any list of sullen malcontents, and there are others with impeccably left-wing pedigrees who shun it like the plague. I cannot believe that proving or disproving that concept could have any bearing on any dimension of modern social relationships, between black and white, male and female, or rich and poor.

I haven't often been in company with other AAT supporters but when it does happen I have found they come from as wide a political spectrum as would be found in, say, a chess club. Indeed John Langdon himself might confirm that, because he once joined a group of us in Ghent, and a genial character he turned out to be. We continued to differ, in perfectly civilised and amicable terms. Categories never came into it.

A final source that might have kept me informed of the counter-case is BBC television. It has an enviable record for its ability to explain complex ideas to the man in the street, assisted by charismatic presenters, interviews with specialists, and stunning film clips. In its flagship science programme *Horizon,* even the most far-out ideas are sooner or later given their day in court. Extra-sensory perception, flying saucers, homeopathy, Big Foot, Chariots of the Gods, the Loch Ness monster, the Da Vinci code - all of them have been investigated by the BBC. The case for them is presented, supporters are invited to state their views, experts come in and pinpoint the flaws in it. By the end of these programmes everyone can understand the conclusion arrived at by the best minds, and exactly how they arrived at it.

It has never taken a look at the water theory. It would be very perverse and ungrateful on my part to complain about that. In 1994 the idea was clearly outlined by Desmond Morris in his BBC series *The Human Animal.* In 1998, the Corporation's Natural History department in Bristol made a 50-minute documentary entitled *The Aquatic Ape* which was sold to and transmitted by the commercial channel *Discovery.* More recently David Attenborough referred to it in his 2002 block-buster series *The Life of Mammals,* in connection with some striking footage of a wading gorilla. He has also presented two half-hour radio programmes about the aquatic idea and the way in which it has been received.

Yet I still regret that *Horizon* never got round to doing one of its polite demolition jobs on AAT, so that we could have heard the arguments on both sides. There must be a strong case against it out there somewhere. I keep hearing it referred to. It would be useful to have it on tape. If I had been the aquasceptic editor of *Horizon,* would I have made the same judgement? Yes, I certainly would. But only if I had discovered that the case against this particular heresy is strangely difficult to put across.

Elaine Morgan

What the fossil record proved

IN THE MINDSET of the PAs, nothing was ever going to make the slightest dent in their convictions that was not proven by the fossil record. And in respect of the Aquatic Theory, as one of its Valkenburg supporters commented: "It is hard to envisage precisely what form such proof might take."

Very hard indeed. Almost all the hominid fossils are of specimens that died by the water's edge, and their bones sank into the silt and were thereby fossilised. In many cases they were accompanied by the remains of fish and crabs and turtles and crocodiles and the odd hippopotamus - but also by occasional land animals. The conclusion drawn from this was that one of our plains-dwelling ancestors had come to the lake or river to drink, and happened to die there. What could be more probable? They had to die somewhere, and the remains of the vast majority assumed to have died on the arid plains would have been eaten by predators or scavengers, and left no trace. It is known as "taphonomic bias." It was and is perfectly true: the fact that their bones sank into the silt can never prove that they lived by the water. But just as certainly, it can never prove that they did not. So that settles nothing.

Then there was the nature of the fossils. Nothing about the fossils of early hominids looked aquatic.

The Naked Darwinist

Again that is incontestable, but what would you look for? It would be just as problematic to dig up the fossil of a mustelid and ask a palaeontologist to determine whether its life-style had been that of an otter or a polecat. Pre-human fossils show clear signs of a shift towards bipedalism, and to me that spells wading. But to the orthodox, it indicates one or other of the assortment of terrestrial explanations for walking on two legs. So altogether it looked like deadlock.

But quite unexpectedly, in the nineties, everything we had all believed about the prehistory of the African continent was turned on his head. Not all of us had accepted that our ancestors had quit the forests and moved out onto the open plains. But as far as I can remember, no-one ever questioned the fact that the open plains were there to be moved onto. Plains are as old as the hills. North America had prairies, Russia had steppes, South America had pampas, and Africa had the savannah. If a primate came down from the trees, it would naturally step out onto the grasslands.

Not necessarily. The PAs themselves were beginning to tell a new story, and part of the impetus came from a new and high-tech sector of bio-historians, the palaeopalynologists. These experts now have the ability to examine fossilised grains of pollen and tell you what kind of plant they would have turned into if they had had the chance to germinate. Other specialists were concentrating on the remains of the smaller creatures which turned up in the same deposits as the australopithecines and other possible forefathers of humanity. All their reports pointed to the same conclusion: the areas that are now savannah were not open plains in those early days. The flora and fauna found in the same deposits as the hominids were not savannah species. At the time when the first human ancestors were walking around on two legs, their environment was covered with trees.

The revelation was very hard to adjust to. When such a firmly rooted collective belief is challenged, the reaction often resembles the one that greeted Darwin's *Origin of Species,* neatly summed up in a Punch cartoon: "If this is true, let us at least hope it will not become generally known." Journalists wanting to make headlines out of the news were strongly discouraged. One reaction caught on film was "Just because the Savannah Theory is wrong, that doesn't mean the Aquatic Theory is right." There was a spate of hasty spin-doctoring. The story was presented as being about climate change. Some climate changes in Africa millions of years ago had simply been re-dated. No big deal. It was helpful that the acquisition of the new facts was cumulative, one little piece of evidence after another. So when journalists got curious about possible wider implications, it was easy to say: "But you can't call this *news!* Where did you get that idea from? No - we've known about this for ages."

Another line was: "This has all been a misunderstanding. There never was a Savannah Theory - that was merely a straw man invented by Elaine Morgan. If anyone did happen to use the term savannah, they did not mean that there weren't trees there, and woodlands, and rivers, and lakes, and maybe some forest. That would have been absurd. Perhaps woodland-savannah would have been a more precise term. But everyone knew what we meant. There's no story here. Nothing has changed."

But it had changed. All the explanatory power of the conventional story was derived from the concept of open grasslands. The hunter chasing the great herds, the heat of the tropical sun, their need of standing up to look over the tall grass and peer into the far horizon, their specialisation in long-distance trekking for mile after mile in the wake of the herbivores - all that thinking depended on the image of the savannah. In woodlands they would not see further through the trees by standing up. There would be no migrating herds to follow. There would be no far horizon to peer into. The long legs would not be needed for speed-running if they were dodging in and out of the trees and the undergrowth. The picture had changed utterly. But the paradigm has not changed. All the adaptations are now explained by pointing out that even if there were trees, they might have been further apart than those in the deep forest where the other apes continued to live. Only think how how hot and sweaty they might have got when moving between one clump of trees and the next. But some of that fossilised pollen was of lianas. You do not find lianas in parklands.

In theory, every scientist should greet joyously every advance in knowledge, whether it confirms or contradicts what he had previously assumed. The PAs had stood up well to this challenge when the geneticists - the new boys on the block - had informed them that their estimate of twenty million years ago for the ape/human split had been quite wrong, and should be amended to five or six million. It was a big jump and it was disconcerting, but after a brief resistance it was accepted with good grace, even though it came from a different group of specialists. In this new case the PAs had every right to congratulate themselves on the new data. Fossil-hunters had made the discoveries, fossil-hunters had published them, and everybody was the wiser. They should have been taking a bow, instead of saying "no, no, this is of no significance". However one of their number, Philip Tobias, did greet it with an upbeat response.

He was the doyen of South African palaeo-anthropology, a disciple of Dart, the discoverer of *Homo habilis,* the custodian of the Taung skull, a zealous life-long promoter of the Savannah Theory. He arrived in London, treating the latest turn of events as a new challenge, an exhilarating opportunity. If orthodox thinking had been on the wrong track, then of course we must go right back to the place where we went wrong, and clearly identify it, and start again from there. But virtually none of his colleagues accepted the invitation to rejoice. They had determined what line to take and Tobias was out of step. They treated him as if he had committed a solecism.

Would I too have tried to play down the problem if I had been embedded in the system? I hope not, but I cannot be sure. The reluctance to say "I was wrong" goes pretty deep in human nature, including mine.

Elaine Morgan

The aftermath

AFTER THE DEMISE of the Savannah Theory, I expected - perhaps naively - that people engaged in the study of human evolution would be compelled to look for something to put in its place. I felt optimistic. My 1994 book *The Descent of the Child* had been well received, perhaps because it touched only briefly on the water theory.

There were other grounds for hope. The rapidity of the dispersal of the hominids out of Africa and across to Asia was leading Chris Stringer to the conviction that they must have migrated around the coasts, rather than taking the more onerous overland route. That of course was no evidence for AAT - it happened very much later than the initial aquatic experience I was suggesting.- but it helped a little to soften the mind-set against any kind of watery influence on our prehistory. And while in England and the United States resistance to the idea remained pretty solid, the books were selling well in Japan, and the Scandinavian countries seemed receptive to it. A Swedish Natural History Museum devoted an exhibition to it, and in Norway a few years later I was awarded the Letten F. Saugstadt prize for a contribution to science.

So I got down to finishing my last book, *The Aquatic Ape Hypothesis*. I tried to put into it every piece of information that anyone had ever tried to follow up in connection with water and evolution, in case the questions they had asked might inspire somebody else. I am told it would have been more effective if I had presented a minimalist case containing only the most irrefutable pieces of evidence. But then I did not think it mattered much any longer, since I was confident that it was the last thing I would ever need to write. The balance of evidence seemed to be moving steadily in our direction. I was immensely encouraged by the fact that a small group of believers in the Aquatic Theory was invited to attend the Dual Congress which assembled hundreds of palaeontologists and biologists from all parts of the world in South Africa in 1998.

So what happened next? Not a lot. Except for one very favourable notice by Chris Knight in the journal of the Royal Anthropological Institute, that last book of mine got short shrift. Reviewers had some kind words for its entertainment value as a good read - "an engaging presentation" - but felt it obligatory to wind up with a curt phrase damning it. The *Nature* piece did it with "Only Morgan acolytes will warm to this appeal." The *Sunday Telegraph* was even more laconic. "What a pity it is probably bunk."

Quite heartening, on reflection. The first quote merely states the obvious: "Acolytes of orthodoxy won't believe it." (No change there, then.) And the second one sounds as if he had first written "What a pity it's bunk", and then went back and put in "probably". Just to be on the safe side.

Officially, nothing has changed. The academic process has continued to operate seamlessly. Questions are addressed, research is conducted, fieldworkers send in reports, genetic analysis plays an increasing role, papers are written and submitted and published. You would hardly notice - indeed, if you were a new student you would not be aware - that the character of the questions has subtly altered. There are more papers than ever dealing with "When?" and "How?" questions, but the question "Why?" - the essentially Darwinian question - is silently by-passed.

Or sometimes not even silently. A paper published in 2007 with the intriguing title of *A new model for the Origin of Bipedality* laid this new policy on the line, pointing out that we are no nearer to agreeing on what benefits were gained by walking on two legs, but we have gathered a lot of new data about genetics. It proposes wasting less time on questions we cannot answer. "Perhaps we need to stop wondering about selective pressures and consider what kind of mutation might be involved …"

"Stop wondering about selective pressures" is a code-phrase for abandoning Darwinism, since selective pressures are the essence of natural selection. The new attitude says "Let's just look at the nuts and bolts, the genes and the chromosomes and the developmental dynamics. If there are questions we cannot answer about why these changes were adaptive, we can stop asking the questions." We can even take a pride in the realism with which we admit our failure, and create a kind of cult of nescience. Another word for that attitude is "hypothesis-free" science, and it represents a huge backward step. Darwin wrote scathingly about the craze for hypothesis-free geology, which even in his day was looked back on as obsolete thinking.

Turning away from Darwinism is made easier by the fact that some of his most fervent defenders are content to make a detour around the unanswered questions. It is possible to mount a brilliant defence of Darwinism in theory, and illustrate it with a wealth of examples from the animal and vegetable kingdoms demonstrating how natural selection accounts for their infinite variety. If there are enough of them, the absence of answers to the "Why only us?" questions may pass unnoticed.

There are two things wrong with this new refusal to ask why. One is that it is based on a mistaken premise. There has not been a failure to find explanations of why we are naked and fat and vocal and bipedal. There has only been a failure to find explanations that the leaders in the field are willing to take a look at. Secondly, these new developments are a gift to creationists. When they demand: "If man was not a special creation, why is he so different?" it is no answer to say: "Oh, didn't we tell you? We don't talk about that any more. We've moved on."

PART TWO

Objections and replies

Objections and replies

THE AIM OF THIS section is to outline the AAT response to some of the stock questions about the physiology of *Homo sapiens*. In it I hope at least to convince you that the mental activity involved is in no way different from, inferior to, or wackier than the orthodox scientific method. It is exactly the same method. It merely starts from a different premise: that a watery habitat may have played a part in shaping us. I know that some readers will be pre-disposed to regard that proposition with scepticism. So before getting down to specifics, it might be worth recapping some of the non-specific reasons for rejecting it that I have heard voiced from time to time.

1. "The palaeontologists have found no confirmation of it."

They have found nothing that proves it and nothing that disproves it. They *have* found things that disprove the Savannah Theory.

2. "The hominids could not have lived by the water because of crocodiles."

That depends on the location. Africa has a long coastline and it has no salt-water crocodiles. Inland, some of its rivers are so teeming with tilapia fish that the crocs are too lazy and well-fed to tackle any more demanding prey. Besides, how would those ancestors have been any safer on land? Imagine a small naked primate in the middle of a plain - we are told that the first ones were no more than four feet high - with no fangs or claws to fight off lions and leopards and packs of hyenas, no night vision to detect the approach of nocturnal predators, and slow-developing helpless young unable to run away.

3. "If our ancestors had been aquatic we would be more stream-lined."

We are in fact far more stream-lined than any other primate. Look at the sleek silhouette of a high diver cleaving the surface of a swimming pool, and try to imagine the silhouette of a gorilla attempting the same manoeuvre. But it takes many millions of years of 100% aquatic life for any mammal to acquire the torpedo-shaped outline of a porpoise. I have never imagined that our ancestor's experience of life in water was on anything like that scale.

4. "Morgan keeps changing her story. In the beginning she said one thing, now she is saying another."

Of course I keep changing my story. So does everybody else in the business, and they would be fools if they did not. In 1972 it was being confidently asserted by all the leading experts that the split between apes and humans occurred twenty million years ago, that Africa in the Pleistocene was in the throes of a horrific drought, and that bipedalism was a consequence of life on the savannah.

5. "The people best qualified to judge are against it."

The people considered best qualified to judge are the people who have spent most of their lives learning and teaching about the current orthodoxy. They are naturally the ones most resistant to change. And they are in a position to ensure that supporters of change find it hard to get promoted or have papers accepted for publication.

7 "I know some people who believe in it and also believe in astrology and ESP and all that stuff."

What does that prove? True, anyone's mind can be too open. (It can also be too closed.) But A. R. Wallace believed in spiritualism, and Isaac Newton dabbled in some distinctly weird forms of alchemy. Those facts do nothing to invalidate Natural Selection or the Law of Gravity.

8. "Some of them have this Eureka moment when they hear of it and don't look for any evidence. That's no way to do science."

True. But I have come across near-apoplectic, knee-jerk reactions *against* it, from people who are proud to say that nothing could possibly induce them to read any books about it. I would reckon these manifestations just about cancel each other out.

9. "The picture she gives has got vaguer as time goes on, instead of clearer."

Not just my vision, but everybody's vision of what happened millions of years ago has grown vaguer as more facts come to light. I remember when David Pilbeam could point out that all the hominid fossil remains discovered up to that date could be contained in a shoe box. There was a general expectation that when there was enough of them, it would become possible, by joining up the dots, to reconstruct the genealogy of *Homo sapiens* with a straight line of "begats", the way St. Matthew traced a family tree from Abraham to Christ. But by now the number of specimens runs into four figures, and the anticipated family tree has turned into something less like a poplar tree and more like a gooseberry bush, with dozens of branches. It appears that at one time there may have been a profusion of different species of anthropoid primates, and most of the lineages led to dead ends.

How has that affected my own thinking? Hardly at all, because one thing I am anxious to clarify is the question I set out to answer. It was, and is, "Why are we so different from the chimpanzee?" The complexity of the gooseberry bush bothers me not at all. The fact remains that, at one point in that proliferating chart, there is one specific intersection, dated - according to estimates which are still occasionally the subject of controversy - somewhere between four and seven or eight million years ago, the point at which the descendants of the LCA - the Last Common Ancestor of chimps and humans - began to diverge into two separate lineages, neither of which petered out. Humans and chimpanzees are both alive and kicking, and the number of ways in which they differ from one another is staggering. It is on a totally different scale from the distinctions between any other pair of species with a comparable chromosomal gap, like lion/tiger, or horse/donkey. Zoom in on that intersection, and ask yourself "Why? What could possibly have happened?"

In search of an answer, let us take a closer look at just a few of the features that strike the innocent eye as distinctive, even though some of them (like the naked skin) have dropped off the current academic agenda.

PART THREE
What makes Human Bodies Special

We walk upright

WALKING ON TWO LEGS is one of the most remarkable of the numerous hallmarks dividing our species from its nearest kin. It may well have been the earliest one to be acquired - it is certainly the earliest of which we have positive evidence - and it has no parallel. No other mammal on land or sea habitually resorts to this erect mode of locomotion.

It is not really surprising that it is rare. Walking the way we do entails a lot of drawbacks. It takes much longer for our children to become independently mobile. Damage to only one limb can cripple us and in the wild that could well be fatal, whereas a quadruped can function very well on three legs. while the fourth one heals. Bipedalism is unstable: when running on uneven ground we may trip and fall. Being erect exposes vulnerable organs to enemy action, whereas in quadrupeds they are safely tucked away. After about five million years of remodelling our skeletons and other organs to specialise in this mode of locomotion, we are still liable to suffer from chronic backaches and other consequential disorders unknown in other mammals. When first resorted to, it must have been more ungainly and far less efficient than it is today. It would have needed some powerful motivation to induce our ancestors to switch to that highly unusual method of getting from A to B.

Elaine Morgan

We can all agree that if it was going to happen to any animal, a primate would be one of the likeliest candidates. Many primates already sit upright in the trees with their spines perpendicular rather than horizontal. Most of them are quite capable of walking on two legs if they feel the urge or see the need to do so, just as easily as we can hop on one leg if we want to. Nevertheless there has to be some good reason why, of all the primates in all the world, only one habitually walks around on its hind legs. The Darwinian question - in the days when we still asked Darwinian questions - was: "In what way did bipedalism prove more advantageous to this one species than to any other?" The answers are familiar but none was ever so convincing that the case was considered closed. Here are some of them.

"On the savannah walking erect enables you to look further into the distance. Meerkats do it all the time: it makes it easier for you to see an approaching predator." (It also makes it easier for the predator to see you.) This is a good reason for *standing* upright, but not for walking or running upright. When the meerkat sees the predator it gets back down on four legs to run away, because running on four legs is much faster than running on two.

"In the wild, chimpanzees on the edges of the forest sometimes stand on two legs to pick fruit." Stand, yes. But they do not move around on two legs, not for more than a few yards.

"Bipedalism frees your hands for other purposes, such as making tools." It may have freed the males' hands, but it enslaved those of the females throughout much of their lives - being upright and naked involved carrying their babies in their arms wherever they went. Chimpanzees sometimes make tools, but just like humans they do it sitting down, not when they are walking around.

"It was a protection against overheating, because if you are upright, a smaller percentage of your body surface is exposed to the perpendicular rays of the mid-day sun." But bipedalism evolved in the woodlands, under the shade of the trees.

"They needed to carry things." What things? Large things, presumably, since chimps readily carry things in one hand when the need arises, and use the other three limbs for walking. The usual answer is "hunter carrying food home to his mate", but this presupposes a settled base from which the hunters sallied forth, and there is no evidence of this until millions of years later than the emergence of bipedalism.

"A man walking upright uses less energy than a chimpanzee walking on all fours." The only reason why we in 2008 use less energy in walking is because we have been practising it and reconstructing our bodies' blueprints for that purpose for at least 5 million years. A chimpanzee on two legs uses just as much energy as a chimpanzee on four legs, whether it

Elaine Morgan

is walking or running. Our ancestors when they first resorted to perpendicular locomotion would have been in the same situation as the chimpanzee, and gained no energetic advantage. Incidentally as soon as we begin to run we still use far more energy than quadrupeds do.

"It was for reaching up to gather the seeds from tall grasses." But they would only have had to tread on them to bring them down to ground level.

"Rearing up on two legs was a signal to other males, to intimidate them. We can see gorillas doing it." Males of many species have a variety of ways of challenging other males, but as a rule it does not pay females to imitate these signals. They are most typical of species with harem-type social systems. Everything about *Homo* - from his vanished canine teeth and the relatively small size difference between the sexes to the size of his testicles - strongly suggests that he is descended from an animal with a less hierarchical way of life.

"We may not be able to outrun a deer, but if we follow it for twenty or thirty miles, we have more stamina and we can tire it out." There seems no intrinsic reason why a quadrupedal animal could not have acquired the stamina to stalk it on all fours. In any case, the early efforts of this biped would have been unsuccessful. Running only nine or ten nine miles would be doomed to failure. An animal does not persist in any type of useless behaviour in the hope that if it perseveres, it or its descendants may one day benefit.

Besides, the earliest bipeds did not run and walk as we do. They used a more graceless gait which the experts call BHBK - bent hip, bent knees. That mode of locomotion is far more energetically costly to sustain than our modern form of erect bipedalism. BHBK is energetically costly for an ape even when it is standing still. It seems strange that our ancestors found it worth while to persist with it for such a very long time before straightening up.

There is just one situation in which all apes and monkeys resort to bipedalism. They all do it whenever they have to cross a stream, or ford a river, or wade into a swamp to gather some of the succulent plants that grow in the water. As Hardy pointed out, if for any reason they waded into the sea or a lake, before the water got very deep they would have no choice but to walk erect if they wished to keep on breathing. Keeping on breathing is more than a powerful incentive - it is an imperative.

A study of captive bonobos showed that they were bipedal for 2% of the time they spent in land, and for more than 90% of the time they spent in water. Algis Kuliukas, who made that observation, is as far as I know the only scientist to have carried out research into the energetics of wading. On becoming interested in the water theory, he abandoned an earlier career, as I had done, "for the subject-sake". But he did it the way I have often been told I should have done it, by enrolling in a university and taking a degree in the subject and becoming a fully paid-up academic in his own right. He has also written a paper which shows that the energy difference between efficient and inefficient gaits is reduced whilst wading, implying that shallow water is a rather ideal place for early hominid bipedalism to have been practised long before anatomical traits evolved which made that bipedalism efficient on land.

For some of today's scientists all the above arguments are out of date. They believe the hard evidence lies in the genes, and Dr. Aaron Filler suggests that the origin of bipedalism was accidental - a purely random mutation in one of the Higher Order Modules in an ancestral primate.

His book is entitled *The Upright Ape: a new Origin of the Species*. He has found evidence for his theory in the fossilised lumbar vertebra of an ape which lived in Africa about 21 million years ago called *Morotopithecus,* caused by a mutated morphogenic Pax gene. The shape of this vertebra would have made it difficult and painful for the animal to have walked on all fours. He suggests that this mutation imposed itself on the lineage that was ancestral to all the extant anthropoid apes - gibbons, orang-utans, gorillas, chimpanzees and humans. He believes that the ancestors of all of them became bipedal. Much later four of them, quite independently and at intervals of millions of years, developed separate modifications of the Moroto lumbar architecture which "allowed them to revisit the proven success of quadrupedal progression." The ancestors of humans for some reason were unable to achieve that. We were stuck with walking on two legs.

The problem is that the original mutation must have occurred in a single individual, and would not have spread throughout the species unless it was adaptive to them in the environment they currently inhabited. The key phrase there is "in the environment." Filler is unspecific about environments. He never mentions the word *savannah* but he does use phrases like returning to the forest. Return from where? And why? We are not told. He implies that all the proto-gibbons and proto-orangs spent formative periods successfully striding around either under the trees or in non-arboreal environments in Asia of which we have no evidence.

This scenario raises far more questions than it answers. It supplies no reason why our ancestors stayed on the ground when the putatively bipedal apes and chimpanzees returned to the branches. True, the Moroto mutation would have been no handicap to becoming, or to remaining, bipedal - but it provided no incentive either. There are many other physical factors quite unconnected with that one vertebra which would make bipedalism a difficult and unlikely mode of locomotion unless there was some strong incentive to resort to it. We need to know what the incentive was. Water is a possible one. Filler does not offer us a better one.

We have a displaced larynx

VICTOR NEGUS DID for the larynx in the twentieth century what Darwin had done for barnacles in the nineteenth. He aimed to search out and record everything that could possibly be known about the subject so that there would be nothing left for anybody else to say. Darwin spent about ten years on the barnacles, and Negus spent a lifetime on the larynx. Their books are available for anyone to consult, but they were so definitive that for a long time, younger scientists hoping to make their way in the world could see no point in returning to these subjects.

The larynx is the upper end of the trachea (the windpipe). In most mammals, it is situated while at rest in the nasal passages above the palate, so that air entering through the nostrils can be conveyed straight to the lungs. Any food or liquid entering the mouth has to pass to one side or other of it before entering the gullet on its way to the stomach. However, when an animal like a dog wants to utter a vocal sound, it temporarily draws down the larynx into the mouth cavity, so that the air and the sound emerge from the mouth instead of the nose; it then goes back up again through a little hole in the palate and is held in place by a sphincter. The same temporary manoeuvre may take place for purposes of panting when the animal is overheated.

In adult humans however, the larynx has moved permanently right down into the neck, below the back of the tongue. It has entirely lost contact with the palate, and never goes back up. This condition has traditionally been regarded as unique to *Homo sapiens*. Darwin was baffled by it because it seems such a bad idea, with nothing to be said for it. Having the openings of the windpipe and the food pipe lying side by side in the bottom of the throat means there is some danger of food particles "going down the wrong way" and getting into the lungs. It explains why people sometimes choke on their own vomit - that could never happen to a dog.

Also, the near-obligatory practice of nose breathing in most mammals ensures that the air is always warmed and moistened and mildly disinfected before it contacts the delicate tissues lining the lungs. Mouth breathing, much commoner in humans, sacrifices this advantage and increases the danger of lung infections. In most animals, mouth-breathing would mean that one of their most important senses - the sense of smell - was temporarily rendered useless, the way sight is when we close our eyes.

Why has this happened to our species? It naturally springs to mind that humans are the only animals with this arrangement and the only ones that can talk: perhaps it emerged so that we would be able to speak? But that makes no more sense than saying that the nose evolved to hold spectacles in place. Nothing

The Naked Darwinist

evolves in order that something else may subsequently happen. Once the larynx had descended it may (or may not) have facilitated or improved our mode of speech, but why did it happen at all? And why only to us?

Here is Negus's explanation. Many of the primates that lived in the trees acquired more erect torsos with forward-facing heads. This change in the angle of the head would be enough in itself to shift the larynx a little way backwards and downwards. That first move, Stage One, begins very early, even before birth, and applies to the infants of all apes and monkeys. And then, in humans, another factor comes into play. In most mammals the tongue lies flat on the floor of the mouth and continues forward into the snout, so there is plenty of room for it. Our own ancestors, like other apes, originally had faces that stuck out in front - but later they became flatter. Negus thought that when that happened, the tongue diminished in size but not in proportion to the reduction in the oral cavity, so the back of the tongue was pushed down into the throat and pushed the larynx down with it, so far down that the epiglottis - the flap at the top of the larynx - lost contact with the palate for good and all. That was Stage Two.

At one point in the last century the larynx became temporarily a hot topic when Edmund Crelin discovered to his surprise (nobody had advised him to read Negus) that human babies are born with the larynx still high in the nasal passage as in other mammals. He noticed that most cases of Sudden Infant Death Syndrome (SIDS or colloquially cot deaths) occur between the ages of three months and six months, the period during which the larynx is on the way down, with its top end not anchored to anything, and capable of flopping around and possibly getting blocked by the uvula if the baby is sleeping in the wrong position. This discovery did not eliminate SIDS, but in Holland between 1987 and 1988, when all the clinics advised parents to put babies to sleep on their backs, there was a 40% drop in the number of SIDS deaths in that country. The advice is now standard in childcare manuals.

Recently interest in the topic has once more been revived by the invention of MRI - magnetic resonance imaging. Scientists can now look at profiles of a man talking or an animal swallowing and see exactly, in real time, just what the larynx is doing. That rendered poor old Negus even more passé, with his Fred Flintstone equipment of scalpel and microscope. Now we could really get somewhere. So what's new?

So far, what's new is not a solution to the problem, but a tendency to suggest that maybe there never was a problem. In Japan a paper was published revealing that the larynx in chimpanzee infants starts to descend in exactly the same way as in human infants. True, but the question is why the process comes to a halt after infancy in chimps but not in humans. From America came the announcement that we are not alone. *Homo sapiens* is not the only land mammal to have a descended larynx. The larynx of an adult male red deer is as least as low in its neck as ours is - lower, in fact. These revelations were felt to be reassuring. A human feature which has puzzled us for 150 years is actually no big deal after all.

I am sorry, I don't get it. That is very interesting about the red deer - so interesting that we must hope to learn much more about it in the near future. Where is the larynx in the female? Is this deer a habitual mouth breather? Does it have a degenerated epiglottis? How has the descent affected its sense of smell? Has it acquired a movable velum like ours, and if so of what size? And what do we have in common with the deer that would cause only these two species out of all creation to acquire this adaptation?

In the case of the deer the reason is convincingly explained. The paper concludes with the suggestion "that laryngeal descent serves to elongate the vocal tract, allowing callers to exaggerate their perceived body size." Newspaper reports added a comment which the paper did not explicitly claim: "This finding raises the possibility that similar 'bluffing' was the original basis for laryngeal descent during human evolution."

That is surely far more debatable. In the deer, the descent occurs at puberty and is sex-linked, but in us it descends in both sexes and in childhood. On reaching puberty, in humans as in many species, it does deepen the voice in males by descending a little bit further than in females. But the larynx of a little girl of six is not designed to allow her to scare the daylights out of her rivals by booming at them. The idea that she has copied this epigamic feature from her father is as unlikely as the idea that she might grow a beard just because he has got one.

You are wondering what any of this can possibly have to do with water. With some trepidation, I am going to disagree with Negus. I do not believe the tongue would have pushed the larynx down. (Sir Arthur Keith did not believe it either.) I think if the tongue needed more room, that was catered for by - and may have been the reason for - the arching of the palate in humans. In apes the palate is flat.

Negus did overlook a few clues. For example, he noted that we are not the only mammals in which the sense of smell is greatly reduced and in which the epiglottis (the flap at the top of the top of the larynx) has degenerated. He named other examples - the sea lion, the walrus, the dugong, and the manatee. He did not draw attention to the fact that those four have something else in common apart from the state of their epiglottis.

Then again, he drew a diagram of the profiles of the heads of twenty-six different species of vertebrates. In twenty-five of them he illustrated by dotted lines the direction of air passing in through the nostrils on its way to the lungs. In one solitary species there was no dotted line, because as Negus commented: "The gannet has no visible nostril". It has no invisible nostril, either. He never asked why, and he never mentioned - perhaps he had never noticed - that some other birds, like cormorants, have nostrils that are visible but non-functional, because they have been blocked off from the inside.

Why? Because they are all diving birds, and they are all obligatory mouth-breathers. Humans are not obligatory mouth-breathers, but we do quite a lot of mouth-breathing. Negus himself pointed out that we can breathe far more efficiently through our mouths than through our nostrils. We automatically switch to that mode whenever we engage in strenuous exercise like running, whereas a race-horse merely flares its nostrils. I think it is worth exploring the possibility that what laryngeal descent primarily facilitates is nothing to do with making sounds - it is mouth-breathing. It greatly enlarges and greatly simplifies the channel by which air is conducted into the lungs.

Incidentally I believe, but have been unable to verify, that we are the only terrestrial species equipped with the gasp-reflex that makes us respond to sudden startlement with a swift intake of breath. It is as if our autonomic reaction to sudden danger was: "Get a lungful of air at once. It might be some time before you can get another." No land animal has evolved to cope with that possibility.

Although Negus repeatedly stressed that man's sense of smell has diminished he never cited other examples of this tendency. All sea mammals become microsmatic. The olfactory bulbs in their brains diminish. In whales they have vanished altogether, and ours are less than half the size of those of apes.

Diving birds and mammals need the ability to inhale large quantities of air very quickly, and for that purpose mouth-breathing is a positive advantage. A descended larynx is one of the best ways of enhancing its efficiency, and would be especially so in our case, since the nasal airways in *Homo sapiens* are possibly the most tortuous and constricted in the entire animal kingdom. The disadvantages of mouth-breathing have often been pointed out: if there had not been some compensating advantage, how would it ever have been selected for? I have been reminded that though in a number of sea mammals the larynx has descended, in whales it has moved in the other direction, upwards. I don't think that disproves my point. It reinforces the idea that when a mammal takes to diving, some changes in the respiratory canal become necessary, while among most land mammals the standard pattern has remained virtually unchanged for upwards of sixty million years.

Ask any Olympian swimmer how much it would cramp his style in a long-distance crawl event, if he had to perform it with his mouth taped shut. I suspect it would adversely affect his performance. It is an idea that could very easily be put to the test.

Elaine Morgan

We can talk

THE LARYNX IS CALLED the "voice box" because, as in other mammals, it contains the vocal chords which enable us to produce sounds. But the evolution of the human ability to speak has very little to do with changes in the position of the larynx. Tecumseh Fitch once reported that while using magnetic resonance imaging he had come across one human subject with the larynx apparently in the undescended position - but no one would have guessed that from the quality of her speech.

The ability to speak has everything to do with changes in breathing. In ordinary breathing, when we breathe in and out, we take in on average about a pint of air and let it out again. Under the influence of strenuous exertion or emotion we may breathe faster or more deeply, but it remains true that inspiration and expiration are equally deep and take an equal amount of time. All this happens without any conscious intention on our part. We can, and do, do it in our sleep. Phonic breathing - the kind needed for speech - is quite different. When we speak we inhale a quart or two of air very rapidly and let it out slowly. We can make it last for 45 seconds or more, if we are preparing to sing an aria or deliver a monologue, and we make a conscious decision to do it.

Only a few mammals are able to exercise this kind of control over their breathing. They are all aquatic. A seal before it dives decides how deep it intends to go and how long it is likely to be before it is able to take its next breath, and acts accordingly. Being able to do that does not necessarily result in acquiring the power of speech - but it is an essential precondition.

The reason you cannot teach an ape to speak is not because its brain cannot grasp the meaning of verbal signals as easily as that of other signals. It can understand spoken commands and obey them. Also, it has nothing to do with any difference in its larynx or its palate or its lips or its pharynx or the shape of its mouth. It is not because it is not willing and eager to learn this trick to please you, or to earn a reward. But it is incapable of doing it. You cannot even teach it to say "Ah" to earn the reward.

Our ancestors did not acquire breath control because it might one day bestow on them a new channel of communication. John Langdon suggested a non-aquatic explanation. In quadrupeds, he said, respiration is locked in phase with gait, but walking on two legs has liberated us from that constraint, since "respiration is independent of locomotion in a biped." True, respiration is locked in phase with gait in some quadrupeds, but that does not apply to the great apes. In species where it is independent of gait, it is instead autonomically controlled by the body's needs of oxygen. Without some compelling reason, being en-

Elaine Morgan

dowed with conscious breath control would be no more liberating than being given conscious control over the bile duct.

It is far more likely that our ancestors acquired it for the same reason that so many aquatic mammals acquired it, because for a life in water it was essential to survival.

We put on weight

The fat layer under our skin was the very first piece of evidence which prompted Alister Hardy to think of comparing the anatomy of aquatic animals with the more enigmatic aspects of human physiology. It led to the observation that while fur is the ideal insulator for land mammals, a naked skin lined with fat may be a better way of keeping warm in water. Researches by P. F. Scholander and others confirmed that it performs that function effectively in aquatic mammals.

On first hearing this suggestion, many people feel it cannot possibly be relevant to compare humans with aquatic animals because many humans are lean, and when they are not it is commonly regarded as purely a consequence of modern life-styles and over-indulgence, and not at all how nature intended us to be. But nature certainly designed us to be different in this respect from other primates. Our babies are born with roughly five times as much fat in relation to body size as is found in apes and monkeys, and for the first year of their lives they continue to grow fatter. *Homo sapiens* is endowed with ten times as many adipocytes - the cells that store fat - as would be expected in an animal of our size.

Today the nations of the developed world are worrying about an epidemic of obesity that is threatening to shorten the average life-span for the first time since advances in medicine and public health began to lengthen it. Other primates, whatever their life-style, are not liable to become obese in the same way because the deposits of fat in their bodies are differently distributed. In primates the characteristic depot is in the paunch, and that is equally true of humans: if we put on weight we tend sooner or later to acquire a fat belly. Nothing unique about that. But it is the fat beneath the skin that makes excessive obesity possible, enabling the fat layer to go on increasing since it is outside the body wall with no bony or muscular constraints on how much it expands.

A book called *The Fats of Life* by Caroline Pond, published in 1998, is one of the most accessible accounts of everything that science has learned about the biological role of lipids in living organisms. It is backed up by an impressive amount of pioneering research carried out by the author at a time when adipose tissue had been widely dismissed as boring and inert and amorphous. It is only in connection with our own species that the author sounds less than convincing.

She describes how different animals store fat in different "depots" - often around the kidneys, or in the abdomen, sometimes at the back of the neck. Animals which need to guard against seasonal scarcity often develop a special site like the camel's hump or the tail of a fat-tailed sheep. But she refuses to refer to the subcutaneous site as a "depot". It is as if its presence there was a kind of aberration, as if fat from more primal and genuine sites has merely happened to migrate, as it were illicitly, to the surface of the body and spread out there to form a continuous layer.

She refuses to endorse the general opinion that the role of the fat layer is to prevent heat loss, maintaining that the fat layer has "nothing to do with thermal insulation." That would certainly spike the water theory, but it would also invalidate most of the conventional speculations. Her opinion is that the fat "may simply be a consequence of our large body mass, as it is in large, obese carnivores such as bears." It is an idea but there are as many or more non-obese large carnivores, and herbivores too.

Even more surprisingly she rejects the use of the word subcutaneous, on the grounds it "implies erroneously" that the fat is firmly attached to the skin. I do not understand this. Whether it is attached or not is not a matter of conjecture but a matter of observation. As far as I can trace, previous anatomists have always described it as attached. Does she disagree? Does her disagreement hinge on the word "firmly"? The point is nowhere developed so it is hard to tell. In short, this is a fascinating book, but I get the feeling that it is trying too hard not to rock the boat.

Elaine Morgan

Stephen Cunnane in his book *Survival of the Fattest*, published in 2005, tends to agree with Caroline Pond in downgrading the importance of the thermo-regulatory function of the fat layer. He stresses the connection between neonatal body fat and the specific metabolic needs of the growing brain which in the final stages of pregnancy constitutes 70% of all the energy needs of the fetus. After birth the infant fat layer - different in composition to adult fat - helps to guarantee a supply of the specific nutrients it needs for healthy growth. Cunnane agrees with Michael Crawford that a shore-based environment is the one most likely to account for this development.

We are furless

A NAKED WOMAN, wrote William Blake, is the work of God. It has proved surprisingly difficult to account for this phenomenon as the work of natural selection. Here are some of the suggestions from earlier years.

(1) Darwin hazarded that our nakedness might be due to sexual selection. But sexual selection usually favours the exaggeration of features which are already typical of the species - if a moth has spotted wings, the females are dazzled by males with the spots artificially enhanced - or signs of health and energy like the elaborate acrobatics of some birds of paradise. But apes are hairy animals, in which the signs of health and nubility are glossy coats, not sparse ones with bare patches. It seems unlikely that the males of one particular group of apes would break ranks and arbitrarily begin to yearn for females with bald bodies, any more than modern males hanker for bald-headed women.

(2) Darwin's contemporary Mr. Belt believed that our ancestors shed their hair because it harboured ectoparasites like fleas and lice, and that their chances of survival would be enhanced by abandoning their body hair as soon as they grew clever enough to light fires and make clothes. This idea has recently been either unearthed or reconceived, and published in a professional journal as a promising contender. But

after millions of years of wearing clothes the ectoparasites that specialise in living on humans are still extant. Besides if we lost our body hair *because* we wore clothes it means that loss of hair came second, so that they must have begun by wearing the clothes on top of the coat of fur. The other way round sounds much more probable.

(3) The brilliant anatomist F. Wood Jones denied that any problem existed, since human hair follicles are actually closer together than a chimpanzee's. Hence "the numerous quaint theories that have been put forward to account for the imagined loss of hair are, mercifully, not needed." He did not confront the question of why so many of our numerous body hairs are so short that they do not appear above the surface of the skin.

(4) A hunting ape on the savannah would get overheated in the chase, and shed its fur in order to cool down. That suggestion was the front runner for decades. But nakedness in land mammals certainly does not correlate with speed, and in tropical latitudes, a coat of fur is as valuable a protection against the sun in the day as it is against the chill of the night. Those who suggest the hominids stood upright to cool down are careful to point to the thatch of hair which has been retained to protect our vulnerable brains from getting overheated. How is it possible that a patch of the same hair that is overheating the body could be so effective at cooling the head?

(5) It has been suggested that in large animals, the ratio of surface area to mass is such that core body temperature can be preserved without expending energy on growing a fur coat - and that we might have dispensed with body hair because we were big enough to do without it. Examples like the elephant and the rhinoceros are quoted in support of this idea. They both belong to a group of mammals Georges Cuvier called pachyderms. Their skin is not only thick and hairless but is characterised by deep wrinkles and/or creases suggesting that their ancestors may have possessed a thick layer of subcutaneous fat. In 1982 my suggestion that the elephant might be ex-aquatic was treated as pure fantasy but recently it has become respectable. If pachyderms are ex-aquatics, the same force that made them naked also made them large, because in water they were functionally weightless and their extra size made no extra demands on their supply of energy. The size theory of nakedness will not work. If animals became naked just because they were big enough to afford to, then gorillas would be even more completely naked than humans, and a buffalo would be as bald as a billiard ball.

(6) William Montagna and his team of researchers, after years of intensive investigation into all aspects of ape and human skin, failed to arrive at any conclusion about the reason for our hair loss. He sadly concluded: "Since it is this single factor which constitutes the chief difference between the skin of humans and the

skin of other mammals, we are left with the major objective of our study still unattained." This is an honest and tenable answer: "We don't know." Presumably those who avoid asking the question are silently concurring with it.

This was the state of play in 2006 when Nina Jablonsky did for skin what Caroline Pond had done for fat. Her book - *Skin: a natural history* - provides a comprehensive and readable account of the evolution of skin, from the integuments of invertebrates through fish, amphibians, reptiles, and mammals, describing its functions and the structures it gives rise to. In any such enterprise, the question of why humans are naked cannot be avoided and she tackles it head on, reminding us that over the last 150 years there has been an abundance of possible answers, "ranging from the well-founded to the wacky." She begins by confronting the aquatic idea, conceding that it is the one with the greatest popular appeal. And creditably, she does not confine herself to uttering the standard formula that it is "not, however, supported by facts." She gives her reasons for rejecting it.

Her reasons include all the usual suspects. Not all aquatic mammals are naked, look out for the crocodiles, humans are not stream-lined, and so on. She adds another that I had not heard of. Assuming that the aquatic venture took place in inland rivers, lakes and waterholes rather than the sea coast, she points out the danger from water-borne parasites transmitting diseases like schistosomiasis, and malaria from the water-breeding mosquito, and the fact that the human immune system does not reflect a history of coping with such parasites.

Admittedly the genetic evidence of an adaptive defence against malaria is found in only a small percentage of the human race and therefore is obviously a recent development. But I am not sure how conclusive that argument is in general terms. Most animals have a long history of being at the mercy of parasites which have been extant almost as long as their hosts, without leaving any detectable genetic evidence of such assaults. The intriguing fact is that three of the ectoparasites which have specialised in living on *Homo* are unable to complete their life cycle unless and until their host enters the water.

Jablonski's critique is clearly and fairly presented. I have only two small quibbles. In the notes she refers to her source of information about the Aquatic Theory as "Elaine Morgan 1982." That early and amateurish effort of mine has long been superseded by one published fifteen years later. And she seems to imply that the alleged advantage to cetaceans of a naked fat-lined skin is simply "reducing drag" rather than being the best form of thermoregulation in water. But the positive virtue of this book is that, unlike many of the aquasceptics, she is prepared to offer her own speculations, which can legitimately be subjected to the same kind of critical analysis.

One conclusion arrived at in this book is that the deciding factor was sweating. It lays great stress on the fact that we are the sweatiest mammals on the planet - that our sweat glands allow us to produce over twelve litres (more than twelve quarts) of eccrine sweat in a day. Jablonski takes the view that once sweating had been selected for, hairlessness had to follow, because sweating would only be effective in the absence of body hair. In fact, that is not necessarily the case. It is not true for example of patas monkeys, which live in open country in equatorial Africa. They are the fastest-running of all the non-human primates, and are capable of prodigious sweating compared to all other monkeys. That seems to keep them adequately cool, despite the fact that they have retained quite a luxurious coat of coarse, shaggy, red-brown hair.

The stress on sweating also seems to militate against some of the other arguments. Any hominids exuding twelve quarts of fluid a day would need to replace it by drinking twelve quarts a day, to avoid collapsing from heat exhaustion. And they would have to drink it slowly and at frequent intervals, because the amount of fluid humans can drink at any one time is unusually small. A very thirsty man can drink no more than 3 percent of his body weight in ten minutes - a thirsty camel can drink 30 percent of its body weight in that time. If they lived far from water they would have had to trek back to the riverside several times every day, or else pay frequent visits to the overcrowded, predator-haunted, parasite-infested water holes. That life-style would not have rendered them immune from contracting either schistosomiasis or malaria.

The thesis here is that we first became sweaty and then had to become naked for the sweat to evaporate. But the arrow of causality would work just as well in the opposite direction: if we first became naked we would have had no protection against the sun's rays and so would need to become sweaty in order to cool down. In either case there remains the "Why only us?" question. Why did this sweaty-and-naked syndrome affect us and not the chimpanzee? The standard answer is the environment.

In Jablonksi's book the environment is not very clearly identified, except that it was away from the water. When the word "savannah" is used it is carefully preceded by "woodland" as if in recognition of the revelations of the nineties. There is no reference to the mighty hunter: these apes are merely described as practising greater exertions, without specifying what the exertions consisted of, or why. There is a passing reference to the least convincing of the savannah's Just So stories - the legend that this primate decided it would be a good idea to stay active under the mid-day sun while all the animals with any sense retreated to the shade.

Elaine Morgan

However Jablonski would have been quite within her rights to have called it "the savannah" loud and clear, because no one knows when the nakedness began - hair does not fossilise - and she has chosen to envisage its onset as occurring much later than bipedalism, at a time when the savannah ecosystem had become well established and the hominid's brain was beginning to expand rapidly. That enables her to introduce a possible "only us" explanation - the fact that the brain is particularly vulnerable to the adverse effects of over-heating. The rapid increase in brain growth happened to our forebears and *not* to those of the chimpanzee. That may have been why the chimp kept its hair and we did not. QED.

It sounds reasonable. I am not carried away by it because there are so many other less costly ways of avoiding an over-heated brain. Jablonski herself lists some of them. One way is to install a cross-current system at the base of the brain which transmits cooled blood from the nose directly to veins at the base of the brain as the blood passes to the heart: this system is used by many mammals like deer, buffalo, and antelope.

Another system is panting, as dogs and numerous other species do. Humans freely resort to panting during exertion to increase our intake of oxygen, and it would have needed comparatively little re-wiring of the nervous system to bring it into play for cooling purposes. A third method is to install a system of veins around the skull that can act like the radiator of a car to keep the brain cool. This one, as Dean Falk pointed out in 1990, humans have already acquired. It might have been enough on its own to cater to the need. Any of these systems, or any combination of them, would have had far less damaging side-effects than losing the protective coat which keeps an animal cool by day and warm by night, protects against abrasions and stings and ultra-violet rays, and in primates provides something for babies to cling to.

In conclusion, the case for an aquatic explanation of our naked skin is not proven: it is hypothetical. But every other explanation on offer is equally hypothetical. New facts about the skin rarely come to light, but last year researchers for the David Attenborough radio programme investigated one question I had raised, namely whether any other mammal secretes the vernix - the cheesy substance that coats the skin of human babies born slightly earlier than full term. They found an example, and it was a seal. That doesn't conclusively prove anything either, but it is certainly interesting.

Most of these questions about human physiology remain unanswered. There are plenty of others if you care to look for them. Why do we have a protruding downward pointing nose? and everted lips? and such a large brain? Why do we shed tears of emotion? Why do we differ from apes in the composition of our blood and the direction of the hair tracts on our bodies? Most of the these questions are never asked, so I have limited myself to highlighting a few which no evolutionist can avoid asking - because they are, as Wood Jones called them, the Hallmarks of Humankind.

It is no longer possible to pretend that the aquatic approach to these questions is wacky and all the others, even if none of them has actually convinced anybody yet, are nevertheless seriously scientific.

PART FOUR
Present and Future

A fairy tale

IF HANS CHRISTIAN ANDERSON were alive today, he might supply a different ending to his tale of the Emperor's new clothes.

When the little boy cried out: "The Emperor has no clothes!", the mounted bodyguard was given the order "Eyes front!" while an equerry reined in his horse and declared "That child is hallucinating." And the crowd? Psychologists have devised numerous tests to show how easy it is to persuade people to doubt the evidence of their own eyes. The crowd looked not at the emperor but at each other to see what other people thought, and the equerry asked them an eminently sensible question, "Which is more likely - that our beloved Emperor is nude, or that the boy is hallucinating?" There is only one answer to that. So the Emperor continued to wear his highly praised garments for the next 35 years.

Elaine Morgan

Where we stand today

MOST OF THE HEAT has gone out of the argument by now. There is a general feeling that the danger - if there was ever any real danger of the aquatic theory gaining acceptance - has passed. The professionals are resigned to the fact that this dissident minority is not going to go away, but they now treat it as only a minor irritant. The latest edition of the *Principles of Human Evolution* devotes a couple of pages to acknowledging its existence, indicating in fairly benign terms that there are well-meaning people around who think in this way, and it is just one of those things. It no longer matters very much.

My own view is that it matters a great deal, because of the effect it is having on the stance adopted by the majority of professionals teaching this subject. It is understandable that since the Dart scenario collapsed they need time to find another one to replace it. It is understandable that many of them nowadays are too absorbed with the exciting new techniques that have been put into their hands to bother about it. This brand of science like any other is subject to what has been called "the tyranny of tools." But some specialists still confront the Darwinian question of why, rather than how, we came to be different. And some of them , as I have tried to demonstrate, are driven to desperate lengths of evasiveness in their anxiety not to be seen to give aid and comfort to the heretics.

It is disingenuous, or worse still self-deluding, to imply: "At the moment the answers to one or two of these questions are not entirely clear." We are not talking about one or two of these questions: we are talking about all of them, all the hallmarks of humankind. They are all unaccounted for. And we are not talking about "at the moment." The answers have been unclear or absent or unconvincing for almost a hundred and fifty years. That strongly suggests that something is missing in the narrative that has been presented to us.

I am sometimes asked: "What is the strongest single argument in favour of the aquatic theory?" There is no single argument. No aquatic explanation of any single anatomical feature is anywhere near conclusive, even where alternative explanations are lame or absent. It is the total picture that makes the conclusion irresistible, because every one of the disparate pieces of evidence points, however tentatively, in the same direction: "Water is one factor that could certainly account for this." In every other paradigm, the different features - the stance, the skin, the voice - tend to be attributed to different causes coincidentally affecting this one single species, or else each is described in circular fashion as a consequence of one or more of the others, or else it is one of the issues people would rather not talk about.

Surely it didn't have to be like this? Is it possible to imagine an alternative course of events in the last century in which this stand-off would never have arisen? There are present-day evolutionists who feel that the phrase "Aquatic ape" presented the wrong image and was a major barrier to acceptance of the idea, and they may be right. Some scientific papers are now beginning to use the word "riparian" to describe the ancestral habitat. Kuliukas thinks in terms of "waterside apes". Verhaegen envisages an "aquarboreal" habitat. Morris thinks in terms of "otter-apes." Stephen Cunnane calls it a "Shore-based scenario." Carl Niemitz uses the language that the profession finds acceptable, describing an "omnivorous, semi-terrestrial quadrupedal locomotor generalist" as the "most probable morpho- and eco-type for an ancestor at the threshold of a hominoid stage of our evolution."

So is it in the end a matter of using the right kind of vocabulary? I strongly suspect if that such a paper, using that kind of language, had appeared in a journal like the *Anthropologischer Anzeiger* in say 1950, it would not have been automatically greeted with hostility or derision as Hardy's articles were. The initials AAT might even have been coined, but used to stand for the Amphibious Ape Theory. After all, the interface between land and water occupies a not inconsiderable portion of the planet's surface. There have always been shorelines, lakes and rivers, estuaries and deltas, floodplains and everglades and intertidal zones, and there have always been plants

The Naked Darwinist

and animals that evolved specifically to exploit them. Amphibians once dominated the animal kingdom, occupying most of the niches later colonised by the reptiles and later still by the mammals. Every single one of them was totally dependent on the co-existence of land and water in its habitat. There is no incompatibility between freshwater and marine scenarios. Even the longest river ends up in the sea, and any riparian ape following it downstream would have found even richer pickings on the coast, and been well placed to initiate the rapid dispersal of our predecessors around the planet.

Fifty years ago, such an alternative AAT might have obtained a respectable place, as one among equals, on the list of possible explanations of bipedalism and other features. The percentage of time spent in water by these amphibious anthropoids could have been envisaged as fluctuating from time to time and from place to place, but as long as they kept one foot on shore, S.J. Gould's query as to why their legs never withered away would not have arisen. Above all, the original idea would have emanated from the right stable. It would never have encountered the unspoken objection that Graham Richards detected in Valkenberg: "If this had been true, one of us would have thought of it."

I am being slowly driven to an unflattering conclusion - that if I had been strangled in my cradle, or had stuck to writing television scripts, this particular branch of study might be in a healthier state than it is today. If so, I regret it. But what's done cannot be undone.

The road ahead

IF YOU EVER GET IT into your head that in one particular respect the leading scientists of your day are on the wrong track, you cannot expect them to welcome that opinion. They will assume you are wrong, and that assumption is likely to be correct. Since they cannot spare the time to explain patiently why you are wrong, it is up to you to dig around, find the reasons why you are wrong, and see the error of your ways.

If no such reasons come to light, you are still in for a long wait. It usually takes at least a generation for new ideas to gain a foothold - long enough for the existing experts (the alpha males) to be succeeded by people who first heard of the heresy while their minds were still comparatively permeable. One cynic observed that such new ideas could only hope to gain acceptance gradually - "funeral by funeral." Three examples will illustrate the factors involved.

The monk Gregor Mendel believed that in sexual reproduction, the hereditary contributions of the two parents did not blend smoothly, like ink and water, as most scientists (including Darwin) then assumed. He thought it was more like mixing sand and sugar - that there were hereditary particles (now called genes) which passed down the generations intact, and obeyed statistical laws. He published evidence for this belief in 1886, but he had no flair for publicity and no eager proselytes to push the idea, so it was around fifty years before his paper was revalued, and longer still before he was acclaimed as the father of genetics.

Raymond Dart, in claiming an African genesis for the human race, was more fortunate. He found an instant convert in Robert Broom, who soon located more fossils to set beside the Taung skull. For many years these claims were ridiculed. But this case was unusual in that no funerals were necessary. It was the two alpha males that had done the rejecting who, after only 23 years, were ready to say "We were wrong and you were right."

In 1929 Alfred Wegener published a book promoting the idea of continental drift, and it was not until thirty years later, after he was dead and gone, that this idea was accepted and incorporated into mainstream science. It is Wegener's name which causes the greatest outrage when mentioned in the same paragraph as my own. It is seen as breathtaking presumption on my part to allow the comparison to be made.

The difference between the two cases, it is pointed out, is that he was manifestly right and I am manifestly wrong. The other difference is that in his case there was a smoking gun which proved him right - the discovery of a mechanism which explained how land masses could move apart. The sudden collapse of savannah theory looked to me like a smoking gun, but the smoke dispersed so quickly that few people were aware of it.

But there are similarities too. Wegener never claimed it was his own idea he was plugging, and neither did I. (Moving continents had been guessed at by Ortelius in 1596, and promoted by F. B. Taylor in 1910.) Another similarity is that for 30 years Wegener's ideas were seen as the essence of crack-pottery - if anything even more so than mine. He travelled to Harvard by invitation to deliver a lecture, but was howled down and had to go home without being heard. If Hardy had gone public in 1940 and compared himself to Wegener, he would not have been accused of presumption. He would have been greeted with gales of laughter. Two of a kind indeed! - that pair of clowns with the bees in their bonnets...

People are still apt to say "Very well, Wegener was right, but for the wrong reasons." His method, if I may venture to say so, was the same as mine. He kept producing little bits of evidence, apparently unconnected, about rocks and plants and animals on different continents, and saying: "Look at this - and this. How do you account for this?" They said "We don't have to account for it." They said: "Go away." But his theory answered all those questions.

There is one last similarity between all these cases. They all got things wrong. Mendel could not understand why his hawkweed experiment went wrong. His second principle of independent assortment was not, as he thought, universally applicable. Dart was wrong about his cherished Man the Killer idea. Wegener was wrong about what might enable continents to move. I have back-pedalled on several issues and now respond to all questions of exactly how and where and why it all started with the safe answer: "I don't know."

Nevertheless, inheritance is particulate, our origins were in Africa, the continents do move, and something caused humans to be different from chimpanzees. Conceivably it might not have had anything to do with water, but no other explanation has turned up or looks at all likely to turn up. At the time of writing the shore-based scenario is the only game in town.

The question is whether there is anything to be learned from all this. For those outside the pale, the lesson is to tread very softly. - which I failed to do - and to have an endless store of patience. For those on the inside, the lesson has already been formulated by Richard Feynman:

"If you get anything new from anyone, anywhere, you welcome it, and you do not argue about why the other person says it is so… You do not have to worry about how long he has studied or why he wants you to listen to him…. We have a way of checking whether an idea is correct or not that has nothing to do with where it came from. We simply test it against observation."

Amen to that.